ONE SMALL SQUARE®

African Savanna

by Donald M. Silver

illustrated by Patricia J. Wynne
and Dianne Ettl

LEARNING
TRIANGLE
PRESS

*Connecting kids, parents, and teachers
through learning*

An imprint of McGraw-Hill

New York San Francisco Washington, D.C. Auckland Bogotá
Caracas Lisbon London Madrid Mexico City Milan
Montreal New Delhi San Juan Singapore
Sydney Tokyo Toronto

Whether you are outside
or at home, always obey
safety rules! Neither the
publisher nor the author
shall be liable for any
damage that may be caused
or any injury sustained as
a result of doing any of the
activities in this book.

Every plant and animal pictured in
this book can be found with its name
on pages 38–43. If you come to a
word you don't know or can't pro-
nounce, look for it on pages 44–47.
The small diagram of a square on
some pages shows the distance above
or below the ground for that section
of the book.

To my parents — *Dianne Gaspas Ettl*

The author is grateful to Penelope Bodry-Sanders
and Karen Malkus for their helpful suggestions;
to Thomas L. Cathey and Maceo Mitchell for
their efforts in making this book possible; and
to Simmie Jooste Grey for a guided tour of her
one small square of Africa.

ISBN 0-07-057931-8
5 6 7 8 9 10 QDB/QDB 15 14 13 12

Introduction

Lions roar. Leopards growl. Cheetahs yelp and jackals howl. Wildebeests grunt, zebras snort, vultures scream, and hyenas whoop. Horns clack, wings flap, hooves thunder, and bones crack. These are the sounds of East Africa. They are the sounds of the hunters and the hunted that live where grass grows as far as the eye can see. They are the sounds of the African savanna.

The African savanna is home to the biggest, the tallest, and the fastest land animals on earth. It is where killer dogs work together as a team, and tiny termites build nests as high as the ceiling in your room and as hard as cement. One minute all is calm. The next, there is panic as thousands of grass eaters run for their lives. The savanna goes from being wet, lush, and green to bare, brown, and dry as a desert. This part of nature is wild and beautiful, full of danger and daring.

As a grassland, the African savanna is mostly flat and open, with only a few trees and shrubs scattered about. More large animals live on the savanna than anywhere else in the world. Beetles, lizards, grasshoppers, and hedgehogs can disappear into the tall grasses. But where do zebras and wildebeests hide from hungry lions and other predators that want to eat them? And how are so many plant eaters able to keep feeding on the same food—grass—without destroying it?

The best way to find out how the African savanna works is to explore it. Not many people are lucky enough to do that. Perhaps one day you will get the chance. Until then, you can explore one small square of African savanna in this book. There will be activities you can do where you live that will provide clues to life on the savanna thousands of miles away.

What's Living on the African Savanna?

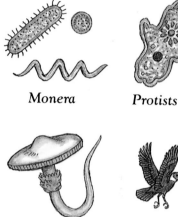

Monera

Protists

Funguses

Animals

Plants

An inexpensive magnifying glass gives you a close-up look at grass growing where you live. That tells you a lot about the grass in Africa.

For some of the activities in this book, you will need binoculars, a mirror, scissors, and a notebook with a pen or a pencil.

An African hedgehog may walk through the small square at night, sniffing for food. It will eat almost anything it can swallow: insects, snails, frogs, snakes, eggs.

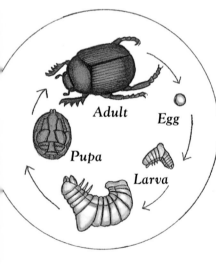

Adult

Egg

Pupa

Larva

Scarab beetles are just some of the insects creeping, crawling, and flying around the square. Beetles change a lot from egg to adult.

One Small Square of African Savanna

Imagine zebras, lions, elephants, and giraffes trying to live in your backyard. Or in a city park. They couldn't. Large animals need a lot of land in order to find plenty of food and water, to hunt, or to flee hungry predators.

The African savanna is a vast space stretching over hundreds of miles. It is the right size for the kinds and numbers of creatures that "make their living" on it.

Just one small square of savanna is shown here. It is 20 feet long and 20 feet wide—about the size of a large living room. This small square is special because it contains both a tree and part of a water hole. Most other squares of savanna the same size would have only grasses.

This square of savanna is full of animals. If you were in Africa, you would never see all these different creatures there at the same time. But if you stayed long enough, you might spot every one of them. That's because different animals use the small square in different ways.

Come explore and discover what the small square holds for the cheetah and the gazelle, the bee-eater and the mongoose. Along the way, life will begin and life will end. There will be feast, famine, and fire—all in one small square.

After sundown, a yellow-winged bat swoops down to catch a grasshopper hiding in the grass. The bat sleeps hanging from a tree branch.

A large male Kori bustard struts around the square as if he owns it. He doesn't scare the carmine bee-eater hitching a ride on his back. The bee-eater catches insects stirred up by the bustard.

Long legs and neck let a giraffe dine more than 17 feet above the square. From that height it can spot a lion sneaking up. Rather than take chances, the giraffe will run to safety.

Need proof that no nest is totally safe? Watch a 6-foot-long boomslang snake make a meal of a weaverbird.

Weaverbirds' nests, woven from grasses, hang from tree branches. Each nest opens at the bottom so it is harder for predators to reach.

Up a Tree

Dark clouds gather above the savanna. Soon sheets of water fall from the sky. The long dry season is over; the rainy season has begun.

Almost overnight, grass sprouts and the small square comes alive again. During the dry months, the umbrella-shaped acacia tree lost its leaves. Then shortly before the first rain, new leaves burst out of buds on branches high above the ground. Now the leaves grow and grow. So do the spiky thorns that keep some animals from eating the leaves.

The sharp thorns don't scare away the tallest animal on earth—the giraffe. A giraffe is a browser. That means it feeds mostly on trees and shrubs. Acacia trees are a

8

Red-billed hornbill

Beware, leaf eaters! Ants live in the thorns of this acacia. To protect their tree, they will attack with a chemical spray. Giraffes don't mind the ants, but elephants cannot stand them.

favorite. A giraffe just lifts its head into an acacia, grasps a branch with its tough lips, and sticks out its tongue to go to work. A giraffe's 18-inch-long tongue can reach about as far as you can with your arm. It can wrap around a branch and pluck off leaves while avoiding thorns. A giraffe may spend only a few minutes feeding at this tree before it moves to another acacia outside the square.

For weaverbirds and hornbills, an acacia is the place to build nests. For bee-eaters, it is the spot to sit and wait for a tasty insect to fly by. To some ants, the tree is home. A leopard will sleep the day away up an acacia. Birds and lizards hunt for insects on branches, while snakes and genets hunt for birds and lizards. A really hungry elephant might just knock over the whole tree to get at the leaves even the giraffes cannot reach.

Shape Up

Are there any umbrella-shaped trees where you live? Visit a park. In your notebook, draw the shapes of some park trees. Look for leaves on the ground and draw their shapes too. How do they compare with acacia leaves? Do any animals eat the leaves of park trees? Surely not giraffes!

Your Savanna Notebook

Whenever you do the activities in this book, keep a notebook and a pen or a pencil close by. Write down what you discover. Draw animals and plants or paste pictures of them in your notebook. Perhaps someday you can visit Africa and record what you see firsthand, as scientists do there.

When an elephant needs to scratch, it may visit the square to rub the itch away on the termite mound. Each mound is a fortress for millions of termites. It contains a chamber for the king and queen, space for funguses to grow, and airshafts, all connected by tunnels. Termite workers build the mound out of soil cemented with their saliva and droppings.

Termites no longer live in this mound. So mongooses, snakes, foxes, and warthogs move in for a night or more.

Funguses grow on termite droppings in the mound. The funguses break down and digest tough plant parts that the termites then eat again.

10

The Tunnel Makers

Just after the first rains, an eagle lands in the small square. It stands beside the tall termite mound and waits. In the acacia tree above, swallows, small hawks, barbets, and other birds also land and wait.

The eagle pokes at the ground and seizes a termite in its beak. In an instant there are more insects than it can handle—thousands and thousands. When the termites take wing, the birds in the tree swoop down to gobble all they can. Most of the termites are eaten, but a few males and females escape. They pair off and start their own nests not far from the small square.

Later in the day, an aardvark makes its way toward the mound. It too wants food. With a slash of its claws, the aardvark opens a gaping hole in the mound. In goes the aardvark's piglike snout and out comes its long, sticky tongue to lap up a mealful of insects. Soldier termites squirt a sticky liquid at the intruder, but that doesn't stop the aardvark. When it finally leaves, blind, wingless workers begin repairing the hole. Meanwhile, other workers are busy digging tunnels and searching the square for food: dead plants.

By tunneling, termites loosen and mix the soil. They also create spaces for air and water that plants and soil animals need to live and grow. In these ways, termites help make sure that countless large and small savanna animals do not go hungry.

King

Soldiers

Queen

Workers

The queen is by far the biggest termite in the mound. She can lay 5,000 to 30,000 eggs a day. Blind workers care for and feed the king and queen. They also build, clean, and repair the mound.

An aardwolf can hear termites at work outside their mound. It quickly laps up all it finds.

Flower

Leaf (blade)

Stem

New grass plant

Roots

Underground stem

New grass plants grow from seeds or underground stems. Each plant can send up more than one stem, producing lots of leaves and flowers. Grasses protect soil from the hot sun, drying winds, and pounding rains. After animals have flattened grasses, the plants usually spring back to life.

Look Out Below!

A hundred miles from the small square, zebras and wildebeests have seen lightning in the distance. Like a flashing neon sign that reads COME AND GET IT, each lightning bolt signals the animals to start at once for the part of the savanna, including the small square, where it is raining. There they will find fresh grasses as the main dish on the menu. And there they will feast for months.

Within a week of the first drops, the small square and the land around it have turned grass-green. By the time the animals arrive at the square, the blades of grass will be growing about an inch a day. But grasses aren't the only

12

You won't find these creatures in your backyard or park. But you will discover similar kinds feeding, hunting, and hiding among grass plants.

thing stirring. Beetles and frogs that spent the dry season asleep in the earth awaken and dig to the surface. Colorful caterpillars and grasshoppers hatch out of eggs and stuff themselves with leaves. Butterflies flit among flowers. Millipedes crawl around stems. Ants raid the termite mound.

Life in the grasses is full of danger. Lurking among the plants are hungry insect eaters: an elephant shrew, a chameleon, a termite frog, or perhaps an African mantis. From above, the beak of a green wood hoopoe may grab an unsuspecting butterfly or beetle. Or the hoof of a giraffe or the foot of an elephant may suddenly squash a patch of grass, sending hidden animals fleeing right into the hungry jaws of a predator.

Growing Grass

Ask a parent or teacher if you may buy and grow grass seeds in an empty pot, can, or milk carton. Or ask if you can work in a patch of grass growing in the yard.

seeds

soil

pebbles

can or milk carton

Look at the plants through your magnifying glass and draw them in your notebook as they grow. Do they look like the grasses on these pages? Do any of the plants flower? Dig up one plant. What are its roots like?

Use scissors to cut different plants at the top, middle, and bottom of the stems. Also trim off leaves in different places. What happens to each cut plant over the next few weeks?

cut   cut   cut

Like your hair, grasses grow
back after being cut.

Grazers have front teeth for biting
and back teeth for grinding. But
only zebras have upper *and* lower
front teeth that can nip the tough
tops off grass plants.

Zebra

Thomson's and Grant's
gazelles both visit the small
square. But "Tommies,"
like this one, come only in
the rainy season.

Thomson's gazelle

Open for Business

Wildebeests leap into the air. Zebras gallop with
delight. Everywhere they look, there are grasses to eat.
There had better be! Hundreds of thousands of animals
are returning in vast herds from their dry-season feeding
grounds. They are grazers—animals that eat mostly grass.

As the grazers arrive, they break up into smaller herds
and spread out over the land. Animals wander in and out
of the small square for a nibble here and a nibble there of
savanna grasses.

Like all other plants, grasses capture energy from the
sun and use it to make sugar. They also make other
nutrients, such as starches, fats, vitamins, and proteins.
When grazers eat grasses, they take in these nutrients to

debeest

A warthog gets down on its knees in order to eat short grasses. Sometimes this wild pig digs up a tasty snack of roots.

use for energy and for growing and staying healthy.

If you watch the grazers, it seems as if they are eating whole plants. They're not: they are feeding on different parts of the grasses. Zebras bite off the tough tops, then wildebeests chew on tender middle leaves and stems. Gazelles nip at the lowest leaves. Instead of harming the grasses, the grazers help them. Cut grasses keep growing back with new stems and more leaves. They also send up new plants from their underground stems.

Day after day, the small square helps feed the herds. Soon it looks like a badly mowed lawn. The grazers sense that it is time to move on. But in a few weeks, when the grasses regrow, they will be back as the square reopens for business.

15

Ambush

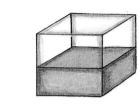

It is almost morning. A zebra leaves her herd and walks slowly toward the small square in search of the safest place she can find to give birth.

When she reaches the square, she sniffs the air for predators. She can't go on—it is time. The birth is quick. Within 15 minutes, the brown-and-white newborn rises on its wobbly feet. It is ready to follow its mother, hungry for her milk.

The sun rises. Soon the heat of day replaces the cool

A lioness's coat color blends in so well with the grasses that she seems to disappear. Can you find where she is hiding?

The world's largest bird stands up to 8 feet tall. It can run at a speed of 45 miles an hour but cannot fly. What is it? An ostrich.

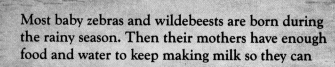

Most baby zebras and wildebeests are born during the rainy season. Then their mothers have enough food and water to keep making milk so they can

of night. All around the square, grazers are feeding. Oxpeckers and other tick birds land on the grazers to breakfast on ticks and bloodsucking flies that annoy the large animals.

To escape the heat, a cheetah rests in the shade of the acacia tree. Nearby her cubs chase each other as if playing tag. Their game helps them practice stalking and pouncing—exactly what these future hunters need to learn. The grazers pay little attention to the cubs but keep away from them just the same.

By late afternoon, many of the grazers are very thirsty. They make their way to the water hole in the square. Not one is aware that female lions have been setting a trap.

As long as the cheetah relaxes on the termite mound, the grazers know they have no reason to fear it.

nurse the young. This zebra is a few weeks old. Its brown stripes will turn black as it grows.

Sound Off

Borrow a video about the savanna from your library or ask a parent if you can rent one. (See page 48 for suggestions. If no savanna videos are available, try any about animals.) Play the video with the sound turned off. As you watch and rewatch it, write down in your notebook what you think the animals are doing. Then view the video with the sound on. Was it easy or hard for you to tell what was going on?

What's It Doing Now?

If you were on the savanna, you would have to stay a safe distance from the wild animals. You would watch them through binoculars or film them to study later. If you have binoculars, take them to the park and try to figure out what different animals are doing. Make notes and drawings in your book. Do any park animals behave like savanna animals? Remember: If you get too close, the animals will flee.

Which Is Which?

Plant eaters have biting and grinding teeth. Meat eaters have long, sharp teeth as well—for stabbing, holding, and tearing into prey. Look at your teeth in a mirror. Can you figure out which ones you use to bite into food? To tear food? To grind food?

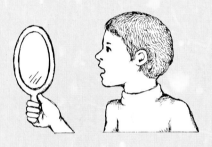

Hitting Stride

An animal that can run faster, farther, or longer than a predator has the best chance of staying alive. The distance an animal covers with each step is its stride. Ask a friend to help you measure your stride when you walk. How does it compare with your friend's walking stride? Watch each other run. At which pace— walking or running—do you cover more distance with each stride? A long running stride can be the key to an animal's escape.

stride

A water hole is a favorite hunting spot for lions trying to catch grazers off guard. Like other predators, lions will target an old, weak, sick, hurt, or slow member of a herd. They will also go after an animal that strays from the rest.

Lions are the most powerful predators on the savanna. They are the only cats that live and hunt together. When lions look at herds, they see plenty of prey. They also see hundreds of eyes, ears, and noses on the alert for the slightest hint of a hungry predator. They see dangerous weapons in the hoofs and horns that can kill them. And they see grazers that can outrun them.

Instead of attacking the herd, female lions split up. Several of them circle the grazers. Then, very slowly, they creep along, bellies to the ground, inching closer and closer to the water hole. Their soft paws hardly make a sound. Meanwhile, the other females choose a spot in the grasses of the small square where the wind will blow their smell away from the grazers. There they crouch as, one by one, zebras and wildebeests come to drink.

Suddenly a gazelle picks up the scent of the circling lions, lifts its head, and stares. An ostrich turns, peers at the same spot, then takes off running. Tick birds squawk, sounding the alarm. Grazers flee in all directions as panic spreads. The circling lions rush the drinking animals, driving them toward the lions hiding in ambush. The trap is sprung. The ambushers leap onto the back of one zebra, sink in their claws, and bring down their prey. The zebra struggles as a lion bites into its throat until it is dead.

When the dust settles, the hungry females gather to fill their stomachs. So do the male lions, even though they rested while the females did all the work.

The cheetah, the fastest land animal, can run at 70 miles an hour. It can catch up to a gazelle fleeing at 45 miles an hour. But the cheetah must kill its prey on the first try. Otherwise, the gazelle escapes because it can run farther without tiring. The young gazelle below won't be so lucky. A martial eagle is about to seize and fly off with it to feed hungry eagle chicks.

Eat in Peace? Not a Chance!

The lions cannot feast on the zebra in peace. Before they even swallow the first taste of meat, vultures are circling high above the square. Vultures are scavengers—they eat dead animals they did not kill. So are jackals, which spot the vultures from a distance. When the vultures dive to land, the jackals know there must be food and head at once for the square.

The lions have no trouble filling their stomachs in spite of the annoying scavengers that keep arriving. The jackals are so hungry that they start to howl. One even dares to steal a scrap of meat from a lioness. Too full to care, she walks off with most of the others to sleep. But the howling has woken a hyena. It too is hungry and

Vultures can spot a dead animal from 1,000 feet up. Their quick arrival on the scene tips off other scavengers. Then the vultures must wait their turn.

Lions live together in groups called prides. The land a pride hunts on is its territory. By marking the territory with a scent that other lions can smell, the pride warns KEEP OUT OR ELSE!

Once most of the lions leave, the spotted hyena makes its move.

20

drives the last lioness away.

Now it is the hyena's turn. Its jaws are powerful enough to crack open bones. No wonder the other scavengers let it feed first. The jackals, however, can hardly wait. They dart in and out, snatching every bite in sight. Finally the squabbling vultures stick their whole heads into the open body and rip out what they can with their sharp beaks. Within hours, little is left. A zebra has lost its life so that many other creatures can keep on living.

On the savanna nothing is wasted, not even animal droppings. Scarab beetles eat them, for they are rich in nutrients. These beetles also roll balls of droppings, lay an egg inside each ball, and bury them. When the egg hatches, the young beetle has food to eat. Uneaten parts of the droppings enrich the soil for growing plants.

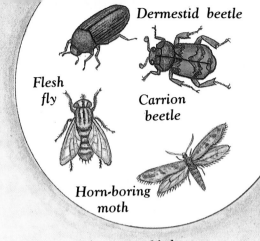

Dermestid beetle

Flesh fly

Carrion beetle

Horn-boring moth

These insects don't mind leftovers. They will feed on anything scavengers miss.

The stork (right) and the raven (below) want all they can get too.

Stealing food from a lion is dangerous. But sometimes it works.

Grazers don't need to watch where they step. Scarab beetles are busy eating and burying droppings as fast as they hit the ground. The beetles help recycle chemicals in the wastes.

What You See . . .

Night hunters depend on their eyes, ears, and noses to find prey. But few see as well at night as during the day.

To see at all, eyes need light. The black dot in the center of each eye is the pupil. It lets light in. Look at your pupils in a mirror in a brightly lit room. Look again after standing in a room with the lights off for 5 minutes. A lion's eye is shown here during the day and at night. Which is which?

Even with wide-open pupils, a hunter has trouble seeing in dim light. Take a piece of red and a piece of blue paper outside just after sunset. Ask a friend to tear a small piece off the red and place it on the blue. Can you find it easily? Try again half an hour later. When an animal's colors and patterns blend into the background, that animal becomes more difficult to spot.

red paper

blue paper

On the Prowl

Night falls, bringing darkness but not safety to the small square. Many grazers are still feeding. Others nod off to sleep for just a few minutes at a time.

From out of the blackness a pack of hunting dogs appears. Slowly at first, the wild dogs walk toward some wildebeests. When they pick up their pace, the wildebeests break into a gallop. The dogs split up into teams that chase one wildebeest from behind and from the sides. In seconds they leap at it. It is theirs.

Hyenas hear the dogs and smell the prey. They patrol the area to make sure no other hyenas invade their territory. Hyenas are scavengers *and* hunters. They could catch their own dinner. But they can't resist attacking the hunting dogs and stealing the meal. So many hyenas want food, however, that they turn on each other and fight for shares.

Hours later, hunger strikes the leopard in the acacia. It climbs down and silently stalks a gazelle in the grasses. After a quick dash, the cat pounces on the smaller animal. Then the leopard drags the gazelle back up the tree, out of reach of hyenas, jackals, and other scavengers. It will be days before the leopard and the lions need to eat again. Predators kill for food, not for fun.

Oops! The genets made a sound that woke the guinea fowl they were after. Say good-bye to dinner!

Hyenas may greet each other by licking and sniffing, and some of their weird night shrieks may sound like laughter. But don't be fooled—they are deadly hunters.

Hungry as they are, adult hunting dogs let their pups eat first. Then they tear apart their prey. The dogs don't frighten a skunklike zorilla about to make a meal of a lizard.

23

On the Way

Giraffes, elephants, leopards, and cheetahs can give birth anytime during the year, even in the dry season. These animals remain when the herds journey to the northwest.

The months pass by. Grazers and predators visit and revisit the small square for food and water. Curious and playful young animals grow quickly under the watchful eyes of adults trying to prevent them from being eaten.

Now there are fewer and fewer rainstorms. The hot sun beats down daily, baking the earth. The grasses stop growing, and their stems and leaves soon toughen. All the animals sense that the dry season is about to begin.

Wildebeests, zebras, and Thomson's gazelles gather. Each evening the water hole is busier and busier. Then early one morning, amid snorts and bellows, the grazers start for the northwest. They are off to river valleys, tall

Dry-season Octavia butterflies are blue. Those of the rainy season are orange. Look for one on page 15.

A hornbill waits for mongooses to stir up insects to eat. The bird isn't lazy. It earns its keep by warning the mongooses of eagles.

24

grasses, and woodlands more than 100 miles away. What a sight as they join tens of thousands of other grazers on the move!

Hyenas and wild dogs follow the herds, ever ready to prey upon them. And the lions, cheetahs, and leopards of the small square have little trouble finding a meal with so many grazers in their territories. But these hunters remain behind after the herds vanish. So do most of the other animals, except for the small birds that fly off to spend the dry season as far north as Europe.

For the time being, there is plenty of food to go around. But all too soon, there will be less and less to eat. Hard times are on the way.

In the rainy season the sphinx moth (1) lays eggs (2) that hatch into caterpillars (3). By dry season the caterpillars dig into the soil and remain pupas (4) until it rains.

What Goes? What Stays?

On the savanna days are warm all year long. There are only two seasons, rainy and dry. When the seasons change, some animals move to new feeding grounds. They return when the seasons change again. This movement is called migration.

Where you live, there may be as many as four seasons. Make a list of animals in your city or town. Which do you see year-round? Which disappear when the weather changes? You can discover where they go by looking in a field guide.

A field guide is a book with the names, pictures, and traits of animals and plants you are likely to find in a given area. A field guide of birds tells you which birds migrate and which don't. You may have field guides at home. If not, you can find them at the library.

The herds face many dangers crossing rivers.
Some rivers have steep, slippery banks. Others
are full of crocodiles. This one has both.

Danger Ahead!

Day and night the grazers that left the square press on. They have no maps or road signs to show them the way. Yet they follow the same routes year after year.

The heat is stifling. Swirling dust, drying winds, biting flies, and hard ground make the journey difficult. Predators make it deadly. They take their toll on the young, the sick, and the old—those unable to keep up.

After days without water, some of the grazers reach a river crossing. For the moment nothing matters but quenching their thirst. There is, however, a price to pay for stopping. Without warning, the jaws of a giant crocodile open out of the water and grab a wildebeest. The weary animal struggles but is pulled into deep water, where it cannot escape.

Crocodiles seem to be everywhere. Even so, the grazers must cross the river or turn back. They jump in, leaping across as fast as they can. Nearly all make it to safety.

With each day, the herds get closer and closer to their new feeding grounds. When they reach the final river crossing, they stampede—gallop as a group. The riverbanks are steep and muddy. Many animals slip, and others run right over them. Before the entire herd has reached the far side, vultures are circling to feed on the unlucky ones. But the rest have made it!

The herds pass by small islands of rocks, called kopjes. Rock hyraxes live there, hiding in holes big enough for them but too small for most predators. The dik-dik in the grass is marking the kopje as part of its territory.

Rock hyrax

Dik-dik

Start your own savanna field guide in your notebook. At the top of a page write the name of an animal or a plant that is in the small square or the dry-season feeding grounds. Write down what you know about it and draw it, using the pictures in this and other books.

Better still, if you can visit a zoo, a wildlife conservation park, or a botanical garden, draw the animal or plant from life. Note features such as size, color, patterns, and sounds. Add to your record whenever you find out something new. (Start fresh pages if you have to.) You can also paste pictures or newspaper articles onto your pages.

Be sure to take your field guide with you if you ever travel to the African savanna.

Not on Vacation

It doesn't take a zebra born in the small square long to notice it is now in a very different place. Instead of a water hole, there is a river full of water to drink. Instead of an occasional acacia tree, there are woods and many bushes nearby. The grasses are much taller than those in the square. And there are animals the young zebra has never seen: impalas, African buffalo, baboons, and even a hippopotamus.

None of these animals is bothered by the zebra nor by the other grazers that keep arriving. There are more than enough tall grasses for so many hungry mouths. The zebras nip off the tough grass tops. Then the wildebeests

Only lions risk attacking buffalo.

The honey guide bird leads the honey badger to a beehive. The badger opens the hive and feasts on honey. Then the bird eats beeswax.

28

Many adult green wood hoopoes feed one hungry baby.

take over, followed by the gazelles.

This is no vacation spot, though. Over the next five or six months, the grazers must keep moving around as they feed, raise their young, and avoid lions and other predators waiting for them to enter their territories. They also must find mates.

Mating time briefly interrupts the peace. Males fight each other over territory, then over females. Most of the time one male backs down before either gets hurt.

Once mating is over, calm returns. By starting new families now, the grazers will be ready to give birth when they return to the small square and the areas nearby during the next rainy season.

Once many black rhinoceroses grazed on the savanna. Today these powerful animals are rare. They are an endangered species.

These impalas can leap 10 feet into the air—but not for a medal. They jump to escape predators.

A plump hippopotamus spends the day with her young in cool river waters. At night they munch on short grasses.

Grass Watch

What happens to the grass plants you are growing when you stop watering them? Do the leaves curl? Do the stems break easily? Cut some at different times and look at them under a magnifying glass.

What happens to the grass in the park or in your yard at the end of summer? At the end of autumn? During the winter? Draw what you see in your notebook.

Cut and Dried

Cut some grass and arrange it on a sheet of paper the way you want it to look when dry. Sandwich the sheet of paper between two layers of waxed paper and stack some heavy books on top. Check every few days to see if the grass has dried onto the paper. Mount the dried grass on a board, cover with plastic wrap, and hang it up to enjoy.

waxed paper

paper

grass plant

Meanwhile, Back at the Square

A lion sits and watches the termite mound where a warthog and her family have moved into a hole left by an aardvark. The lion finds warthogs tasty all year long, but especially during the dry season.

With most grazers gone, lions and other predators have to change their diets. A Grant's gazelle makes a meal but is not as filling as a zebra or a wildebeest. Attacking a young giraffe is risky if the mother is near. Should all else fail, predators turn into scavengers and look for vultures to lead them to food.

African bullfrog

Hedgehog

Termite frog

The mongooses better hurry underground to escape the eagle.

Some animals don't have to leave the square to survive the dry season. The African bullfrog buries itself in the mud at the bottom of the water hole. The termite frog digs into the mound, and the hedgehog into the ground. Neither moving nor eating, they sleep until the rainy weather returns.

Plants must also change their way of life to stay alive without rain. When there was lots of water, the plants soaked up more than they needed. From tiny openings in their leaves, they gave off what they didn't use. Now water is far too precious to let any escape. So the acacia sheds its leaves. Grass blades stiffen, turn brown, curl, and die. Both the acacia and the underground parts of the grasses live off stored food.

There is still some water to drink, but the supply is slowly shrinking. That is of no concern to the frogs, toads, and hedgehogs that have buried themselves in the ground. They sleep there until the rains return.

Don't mess with a mother warthog. To protect her young, she will attack a lion, a leopard—even an elephant.

Grant's gazelles

Bacteria and other very tiny decayers break down dead grasses and recycle minerals back to the soil.

Protozoan

Bacteria

A giraffe can go almost a week without water. Soon it may have to.

31

Desperate for water, one elephant digs into the dry ground. Another peels bark to get at water stored inside a tree.

Don't Call 911

There has been no rain for months. The heat doesn't let up. The ground cracks in places. Many animals die and become food for those still able to go on. An elephant crosses the square and digs down with its tusks into the dried-up water hole. It is searching for water, but there is none to be found. So it moves on.

Later that afternoon, lightning streaks through the sky. It is a false alarm, for not a drop of rain falls. However, one lightning bolt strikes a patch of grasses, setting it on fire.

In seconds, crackling flames rise from the earth. The wind fans the blaze, spreading sparks that reach the square. It too goes up in flames. The grasses burn quickly. Neither the termite mound nor the acacia catches fire. And all the underground plant parts are likewise unharmed.

There is no need for fire trucks to put out this wildfire. In nature, fire clears the land of dead stems and leaves and destroys young trees trying to invade grasslands.

Fire also means food. Standing at the edge of the burning grasses are large birds eager to swallow snakes, lizards, and insects fleeing the flames.

When the smoke clears, a layer of black ash covers the square. It will enrich the soil with nutrients from the burned plants. The square is bare—but not for long!

When there is dryness or fire, grass seeds (1) twist down and plant themselves in the ground (2). They remain there, unharmed (3), until new rains signal them to sprout.

Seed
1
2
3

Secretary bird

No need to stomp through the grasses today after food. The secretary bird picks up its meal slithering out of the fire.

33

After the long, hot, dry season, life-giving water falls on the small square and the savanna around it.

When the fire passes, the ground is blackened. The midday heat seems stronger than ever. Yet the acacia bursts into flower before any rain falls.

Adult running frog

Eggs

Tadpoles

When the rainy season starts, this frog lays eggs in puddles. The eggs hatch into tadpoles that swim and breathe underwater through gills. When the tadpoles lose their gills and tail, they grow lungs and legs. They change into adult frogs that live on land. The adults sleep away the dry season buried in the mud.

34

Getting the Job Done

Without a weather report, the acacia starts to bloom again. The grasses are ready to grow. The animals of the small square need water. Then one day the air feels thick. Flat-bottomed clouds build higher and higher. The sky darkens. Lightning flashes. Thunder rolls. A cool wind sweeps over the savanna. This is it!

The first drops soak into the thirsty earth. That's all the grasses need to start growing again. The dry season is finally over. The rainy season has begun. And you can bet that zebras, wildebeests, and gazelles will soon be on their way back to green pastures.

Year after year nature does the job of feeding many creatures without harming the savanna itself. Nature depends on everything living on the savanna to do its

Atop the termite mound the cheetah stares into the distance. Is she already on the lookout for the herds from the north?

35

Guessing Game

Draw a circle about 10 inches in diameter on each of 2 sheets of paper. Cut a pair of windows out of one circle as shown. Divide the other circle into 6 pie pieces. Number the pieces 1 to 6 and draw the following:

1 – lion
2 – wildebeest
3 – eagle
4 – zebra
5 – hunting dog
6 – gazelle

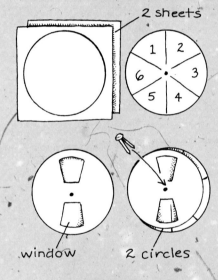

2 sheets

window   2 circles

Place the "window" circle on top of the "pie piece" circle and fasten them together with a paper fastener through the center. Have your family and friends turn the wheel and see if they can figure out which window contains the predator, which the prey. Create more circles using other animals in this book.

Savanna Diorama

Take a shoebox and measure its length and height. Cut a piece of paper for the background wall about ¼ inch less high than the box but about 4 inches longer. On it, draw and color grasses, an acacia tree, and the sky. Place the picture in the shoebox and tape each side to the front. The picture will curve.

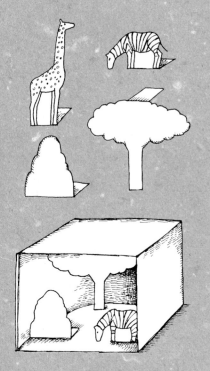

On separate sheets draw and color more grasses, a termite mound, and different animals, each with a flap at the bottom. Cut each out, bend its flap, and glue or tape it to the bottom of the box. Hide some animals in the grasses. You can also draw and color part of a water hole on the bottom of the box.

part to keep the savanna working the way it should. And that includes everything in the small square.

The plants are nature's food makers. Grazers are gardeners that trim grasses. They are food preparers that help each other get at the parts of grasses they like best to eat. Plant eaters are also protectors when they warn whole herds of danger.

Predators, too, do their share. They help keep the number of plant eaters from exploding. Too many grazers, for instance, would quickly wipe out all the grasses. Instead, they move on and give the plants a chance to regrow.

Scavengers and decayers make sure nothing is wasted. Scarab beetles enrich the soil with buried droppings. Termites let air and water into the soil by tunneling. And recyclers return minerals and other nutrients to the earth for plants to reuse.

Nature can continue getting the job done as long as savanna animals have all the space they need and the grasses stay healthy. But farmers and ranchers have started to take over savanna land. They replace zebras and wildebeests with cattle and sheep. Cattle and sheep are grazers too. But unlike savanna animals, they eat grasses down to the bare ground, kill roots, and damage the soil. That's why parts of the savanna have been set aside to protect them. But a lot more needs to be saved—not only for the plants and animals, but for nature lovers like you to visit and see life at its wildest.

Can you match each outline to its picture in the small square?

Martial eagle

Masai giraffe

Thomson's gaze

Cattle egret

Droppings w beetle eggs

Leopard tortoise

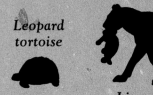

Lion

White-naped raven

Crested guinea fowl

Carmine bee-eaters

Nest

Village weaverbird

Crested hoopoe

Verreaux's owl

Acacia tree

Widowbird

Yellow-winged bat

Leopard

Red-billed hornbill

Eland

Small spotted genet

Wildebeest

Kori bustard

Green wood hoopoe

Spectacled elephant shrew

Termite mound

Burchell's zebra

Ruppell's vulture

Termite frog

African bullfrog tadpoles

Agama lizard

Dwarf mongoose

Waterbuck

Kenya mole rat

African rock python

Grass

Cheetah

Secretary bird

Bat-eared fox

Sacred ibis

37

Most of the predators and grazers on the savanna are mammals. Mammals are vertebrates—they have a back-bone. They make their own body heat. They are the only animals that grow hair or fur and make milk for their young. All these animals are mammals. You are a mammal too.

Mammals

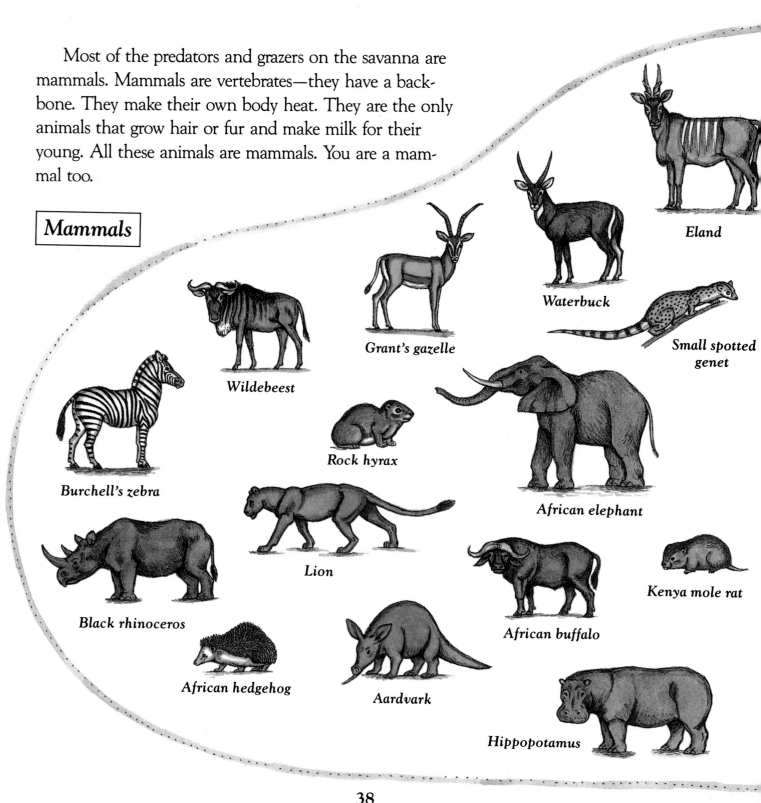

Eland

Waterbuck

Grant's gazelle

Small spotted genet

Wildebeest

Burchell's zebra

Rock hyrax

African elephant

Lion

Kenya mole rat

Black rhinoceros

African buffalo

African hedgehog

Aardvark

Hippopotamus

Impala

Thomson's gazelle

Spotted hyena

Leopard

Aardwolf

Black-backed
jackal

Bat-eared fox

Warthog

Masai giraffe

Baboon

Cheetah

Dik-dik

Wild dog

Springhare

Yellow-winged
bat

Spectacled
elephant shrew

Ratel

Dwarf mongoose

Zorilla

Like mammals, birds have a backbone and make their own body heat. But birds are the only animals with feathers. Every bird grows them—even ostriches, which cannot fly. Instead of teeth and heavy jaws, birds have tough, light, toothless beaks.

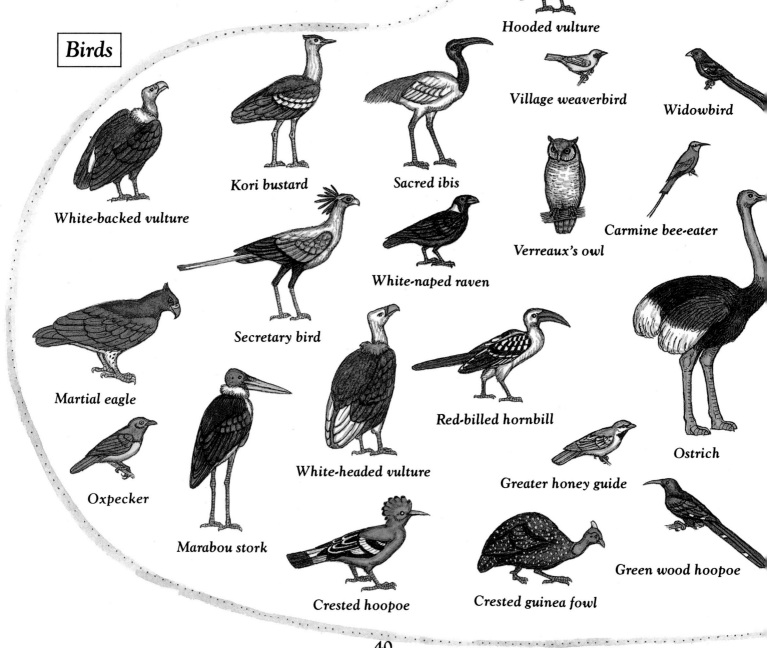

Birds

Ruppell's vulture

Hooded vulture

Village weaverbird

Widowbird

White-backed vulture

Kori bustard

Sacred ibis

Verreaux's owl

Carmine bee-eater

Secretary bird

White-naped raven

Martial eagle

Red-billed hornbill

Ostrich

Oxpecker

White-headed vulture

Greater honey guide

Marabou stork

Crested hoopoe

Crested guinea fowl

Green wood hoopoe

Reptiles and amphibians are also vertebrates—animals with backbones. But reptiles and amphibians cannot make their own heat. They turn as warm or as cold as the world around them. Most reptiles have a thick skin with tough, dry scales or bony plates.

Reptiles

Cordon-bleu waxbill

Nile crocodile

Boomslang snake

Leopard tortoise

Jackson's chameleon

Monitor lizard

Agama lizard

African rock python

Great white egret

Little egret

Cattle egret

Amphibians live the first part of their lives in water and the rest on land. They have smooth, moist skin without scales.

Amphibians

Termite frog

Running frog

African bullfrog

41

These animals are invertebrates—they have no bones. Most of the invertebrates in the small square are insects. Adult insects have six legs, a hard crust covering the outside of their bodies, and usually wings. Young insects, such as caterpillars, can have many legs. Like all other invertebrates, insects cannot make their own body heat.

Insects and Other Invertebrates

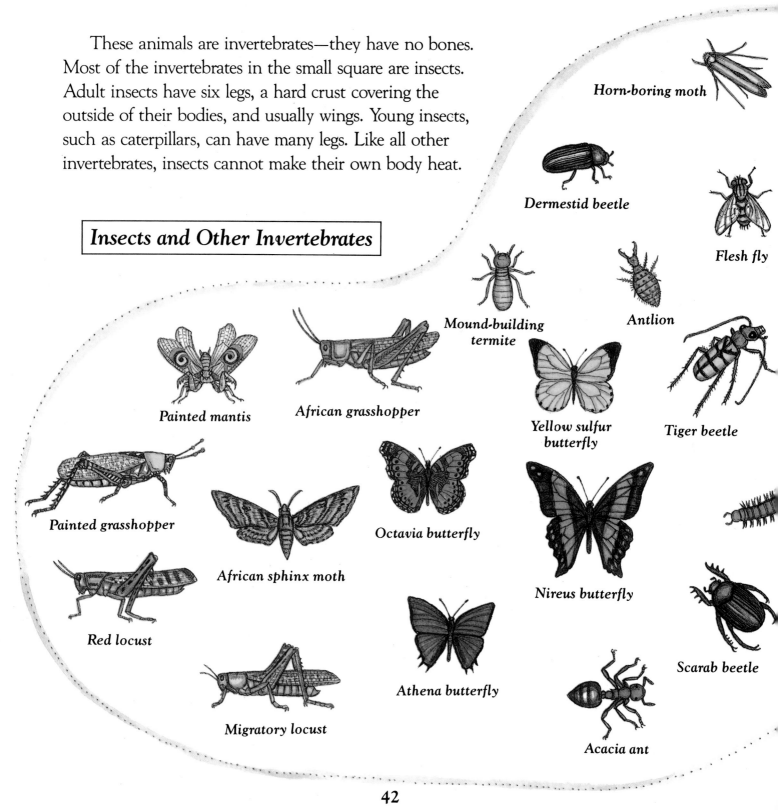

Horn-boring moth

Dermestid beetle

Flesh fly

Mound-building termite

Antlion

Painted mantis

African grasshopper

Yellow sulfur butterfly

Tiger beetle

Painted grasshopper

African sphinx moth

Octavia butterfly

Nireus butterfly

Red locust

Athena butterfly

Scarab beetle

Migratory locust

Acacia ant

The most important plants growing on the savanna are grasses. They survive heat, drought, fire, and being grazed and stepped on. Grasses and other plants make food using energy from the sun. Funguses, which often look like plants, cannot make food.

African honeybee

Scorpion

Carrion beetle

Millipede

Plants

Thorn acacia

Umbrella acacia

African daisy

Glory lily

Aloe

Elephant grass

Red oat grass

Bermuda grass

Fungus

Canary grass

June grass

Bluestem grass

Termite-mound fungus

Monera

Protists

Ameba

Cellulose-digesting protozoa

Bacteria

Protists and monera can be seen only under a microscope.

43

Index

A

aardvark (AHRD-vahrk) 11, 30
aardwolf 11
acacia (uh-KAY-shuh) 8, 9, 11,
    17, 22, 28, 31, 34, 35, 36

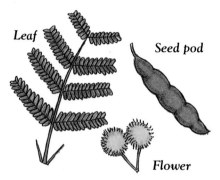
Leaf   *Seed pod*

Flower

amphibian (am-FIB-ee-in) 41
ant 9, 13

B

baboon 28
bacteria 31. *Kinds of monera—*
    one-celled creatures that don't
    have a nucleus, a control
    center.
barbet 11

bark 32
bat 7
beak 11, 40
bee-eater 6, 7, 9

beehive 28
beetle 5, 6, 13, 21, 36
binoculars 5, 17
bird 9, 11, 25, 32, 40
bone 3, 21, 42
boomslang 8
browser 8
buffalo 28
bustard 7
butterfly 13, 24

C

caterpillar 13, 25, 42
chameleon (kuh-MEEL-yin) 13
cheetah 3, 6, 17, 19, 24, 25, 35
claw 11

crocodile 26, 27

D

decayer 31, 36. *Living thing that*
    breaks down dead plants and
    animals.
dik-dik 27
droppings 10, 21, 36
drought (drowt) 43. *A long*
    period with little or no rain.
dry season 8, 14, 24–33, 34, 35

E

eagle 11, 19, 24, 30, 35
egg 6, 13, 21, 34

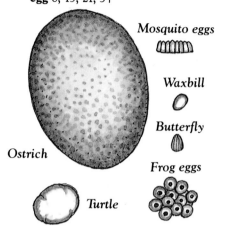
Mosquito eggs

Waxbill

Butterfly

Ostrich   *Frog eggs*

Turtle

elephant 6, 9, 10, 13, 24, 31, 32
elephant shrew 13
endangered species 29
energy 14, 43. *Ability to do work*
    or to cause changes.

F

fat 14
field guide 25, 28
fire 6, 32, 33, 34
flower 12, 13, 34

Index

fly 17, 27

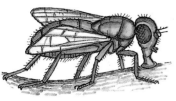

fox 10
frog 6, 13, 30, 31, 34
fungus 5, 10, 43

G
gazelle 6, 14, 15, 19, 22, 24, 28, 30, 35
genet 9, 23
gills 34. *Breathing parts of tadpoles, fishes, and many other water animals.*

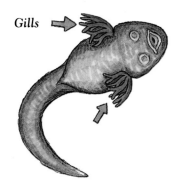

Gills

giraffe 6, 8, 9, 13, 24, 30, 31
grass 3, 5, 6, 8, 12, 13, 14, 15, 16, 22, 24, 25, 28, 29, 31, 32, 33, 35, 36, 43
grasshopper 7, 13
grassland 5, 32
grazer 14, 15, 17, 18, 19, 21, 22, 24, 25, 27, 28, 29, 30, 35, 36, 38
guinea fowl 23

H
hawk 11
hedgehog 6, 30, 31
herd 14, 15, 16, 18, 19, 25, 26, 27, 35
hippopotamus 28, 29
honey badger 28
honey guide 28
hoof 3, 19
horn 3, 19

hornbill 9, 24
hunter 3, 6, 17, 22, 23, 25
hyena 3, 20, 21, 22, 23, 25

I
impala (im-PAL-uh) 28, 29
insect 6, 7, 9, 11, 21, 24, 32, 42
invertebrate 42

J
jackal 3, 20, 21, 22

K
kopje (KAHP-ee) 27

L
larva (LAHR-vuh) 6. *Very young animal that doesn't look at all like its parents.*

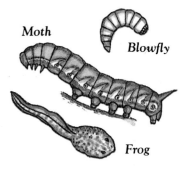

Moth

Blowfly

Frog

leaf 8, 9, 12, 13, 15, 24, 31, 32
leopard 3, 9, 22, 24, 25, 31
lightning 12, 32, 35
lion 3, 5, 6, 8, 16, 17, 18, 19, 20, 21, 22, 25, 28, 29, 30, 31, 35
lizard 5, 9, 23, 32
lungs 34

M
magnifying glass 5, 13, 30
mammal 38, 40
mantis 13
mating 29
migration 25
milk 16, 17, 38
millipede (MIL-uh-peed) 13

Index

mineral 31, 36

monera (muh-NEER-uh) 5, 43. *Creatures made up of one cell that doesn't have a nucleus, a control center.*

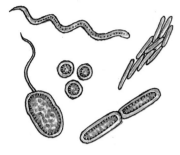

mongoose 6, 10, 24, 30

moth 25

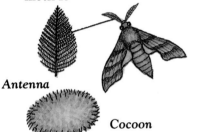

Antenna

Cocoon

mud 30

N

nest 3, 8, 9

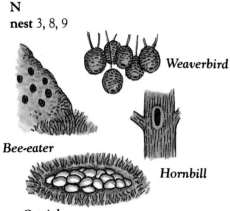

Weaverbird

Bee-eater

Hornbill

Ostrich

notebook 5, 9, 13, 17, 28

nutrient (NOO-tree-int) 14, 21, 36. *Any part of food that living things must have to build cells or to use as a source of energy.*

O

ostrich 16, 19, 40

oxpecker 17

P

plant 5, 10, 12, 15, 21, 31, 36, 43

plant eater 5, 35

predator (PRED-uh-tur) 5, 6, 8, 13, 16, 18, 19, 22, 24, 27, 29, 30, 35, 36, 38

Marabou stork   *Monitor lizard*

Mongoose

prey 19, 23, 35. *An animal hunted or caught for food by a predator.*

pride 20

protein 14

protist 5, 43. *Creature usually made of one cell that has a nucleus, a control center.*

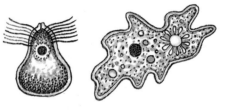

pupa (PYOO-puh) 6, 25. *Part of some insects' life cycle when their bodies change completely as they become adults.*

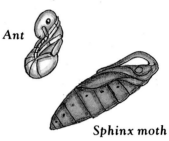

Ant

Sphinx moth

Horn-boring moth

R

rain 11, 12, 24, 25, 30, 31, 32, 33, 34

rainy season 8–24, 25, 29, 34, 35

raven 21

recycle 21, 36. *To use over and over again.*

reptile 41

rhinoceros 29

river 25, 26, 27, 28, 29

rock hyrax 27

root 13, 15, 36

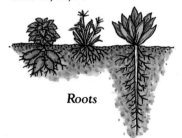

Roots

S

saliva 10

Index

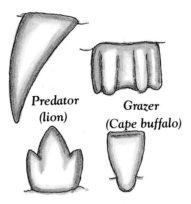

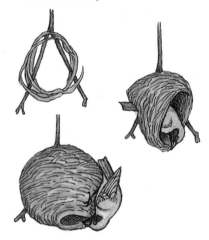

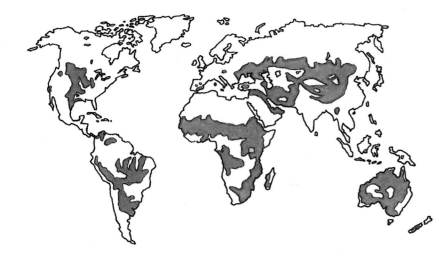

The gold area on the map is the savanna in East Africa, where the plants and animals in this book live, called the Serengeti. The green areas show where other grasslands are found around the world. Most are in the middle of continents.

Nature Videos

Look for the following in a library or video rental store:

National Geographic Videos African Wildlife, Lions of the African Night

Time-Life Video The Living Planet; Sea of Grass; Trials of Life; Elephant, Lord of the Jungle

PBS Home Video Nature: Leopard, Darkness in the Grass

Armchair Safaris Africa, Serengeti Migration

Nova Video Library Animal Olympians

Audubon Video On the Edge of Extinction

Animals of Africa Volumes 1–9

Serengeti No Place to Hide, Parts 1 and 2

Life on Earth

ABC World of Discovery Videos